Green Gone Wrong

Green Gone Wrong

Ecopolitics Exposed

Paul Taylor

Writers Club Press
San Jose New York Lincoln Shanghai

Green Gone Wrong

Ecopolitics Exposed

Writers Club Press

an imprint of iUniverse.com, Inc.

For information address:

iUniverse.com, Inc.

5220 S 16th, Ste. 200

Lincoln, NE 68512

www.iuniverse.com

(800) 376-1736

ISBN: 0-595-16161-8

Printed in the United States of America

To Dr. and Mrs. A. N. Taylor—

who allowed me to make my own mistakes, and discover that a long and prosperous life must be lived in the balance of privilege and responsibility.

Contents

Preface

This work of nonfiction is an overview assessment of the environmental movement that began, and reached its full legitimate expression, in the United States during the 20th Century.

The term *environment* essentially means *surroundings* (36). Unfortunately, its meaning has been stretched, conflated, contorted, misappropriated, abused, bastardized and politicized to the point of trivia by activists and media to manipulate public policy in a range of issues that are as endless in their scope as often misguided in their ends. In the US, environmentalists have become skilled at gaming the environmental regulatory system for political advantage in the guise of politically-correct progressive public service as tax-exempt organizations. This is today's *ecopolitics*. Predictably, environmentalists identify the sinister enemies of the environment as greedy corporations, property owners and affluent lifestyles. It starts with media access to promote civic fear and misapprehension about environmental issues. This is the *Alice-in-Wonderland* area where the possible becomes the probable, the suggestive becomes the conclusive, the accusation becomes evidence.

Today, computer models can generate compelling data for **any** political argument on an untestable proposition about a future hypothetical environmental threat, cloaked in a veneer of pseudo-science. Pseudo-science often sustains positions of environmental movements where a tenuous tissue of worst-case assumptions requires the rational investigator to prove the irrational negative proposition. Here, the obscure, often immeasurable environmental impact is promoted as an imminent threat until the contrary is proven. This *fear mongering* is a disingenuous tactic to erect counterfeit public issues for political exploitation.

The *political left* has no interest in environmental successes because their issue identity and ethos are lost in the recognition of the successes. The *political right* has no interest in environmental successes because it would require the recognition that some government environmental programs have worked. As with most public issues, the long-term practical solutions derive from moderate centrist policies that do not attract activists, media or political extremes, or their inherent powers of public persuasion and fund raising.

The reader is invited to consume this work with a mind open to the new reality that the US environmental movement has lost its way in the face of its own seldom-promoted successes in the 21st Century.

Acknowledgement

The publication of this book has reasonably applied *best efforts* in citing and referencing sources consulted in the preparation of this book pursuant to the *fair use doctrine*. The author wishes to express his sincere gratitude to the sources consulted in adding to the body of public knowledge on subject matter as important and intractable as *the environment*.

Disclaimer

This book is written to provide information on the subject matter herein. It is published and sold with the understanding that the publisher and author are not offering legal, scientific or business advice on any specific project, entity, person(s) or circumstance. A *good faith effort* has been made to publish this book accurately and completely, therefore any typographical, or content errors or omissions are inadvertent. Neither the author nor publisher of this book shall be liable, responsible or party to any claimant(s) to loss or damage, caused or alleged, directly or indirectly by the purchase, contents, use or inference of this book.

Abbreviations

AIDS   Acquired Immuno-Deficiency Syndrome

CERCLA   Comprehensive Environmental Response, Compensation and Liability Act.

CFC   Chlorofluorocarbon

CFR   US Code of Federal Regulations

CO   Carbon Monoxide

CWA   Federal Clean Water Act

DDT   Dichlorodiphenyltrichloroethane

EPA   US Environmental Protection Agency

ESA   Federal Endangered Species Act

FBI   Federal Bureau of Investigation

FR   Federal Register

FWS   US Fish and Wildlife Service

GAO   US General Accounting Office

GDP   Gross Domestic Product

IRS   US Internal Revenue Service

LUST   Leaking Underground Storage Tank

MCLG   Maximum Contaminant Level Goals

MTBE	Methyl Tertiary-Butyl Ether
NEPA	National Environmental Policy Act
NIMBY	"Not in My Backyard"
PAC	Political Action Committee
RCRA	Resource Conservation and Recovery Act
R.E.A.	Registered Environmental Assessor
RFG	Reformulated Gasoline
SARA	Superfund Amendments and Reauthorization Act
SDWA	Federal Safe Drinking Water Act
SO_2	Sulfur Dioxide
TMDL	Total Maximum Daily Load
T.R.	President Theodore (Teddy) Roosevelt (1858-1919)
US	United States of America
UST	Underground Storage Tank

Introduction

Few, if any, public issues have the personal resonance of environment issues. Each of us can feel both victim and perpetrator in environmental issues. Environmental issues have become an integral part of American culture. The environmental movement has grown worldwide to become the largest, most densely organized political cause in human history (46). The US environmental regulatory system, begun in earnest in the 1960s and 70s, has often erred on the side of caution under the paternalistic maxim of "prudent avoidance," in the absence of complete data on actual impacts, to now control everything from the propellant in your deodorant, to your backyard barbecue fuel, to incidental drainage ditches on private property, to subspecies of plants and animals that serve only as curiosities in ecological function.

From a broader perspective, without a clear scientific consensus on the validity of such things as *global warming*, activists have hijacked political agendas in yet another doomsday scenario that if true is the most apocalyptic so far. However, if false, the trendy global warming hysteria may be the final blow in the precipitous slide in legitimacy of environmental activism. Ironically, in the 1960s and 70s, environmentalists successfully defeated the expansion of nuclear power plants in the US that could have substantially reduced greenhouse-gas-producing fossil fuel use impacts which those environmentalists now blame for global warming. Nuclear power plants also reduce US dependence upon foreign oil. Today, nuclear power only accounts for 19% of the US electric power supply (26). There is a profound arrogance in the belief, without firm scientific proof, that human activity can significantly impact global climatic conditions. 95%

of the sources of the predominant global warming gas carbon dioxide, which comprises only 0.03% of our atmosphere, are naturally occurring sources (3).

Neither politics nor the legal system has consistently been accountable to the truths of environmental science. It is now time to impose a moratorium on any further US government environmental regulations until true science is re-introduced into environmental policy decisions. Government regulations of any kind are the specific implementations of the general legislative intent of laws with the effect of controlling private and public activity by limiting liberty and freedom.

When dealing with attitudes toward the earth's ecosystem, we must recognize the absence of a commonly accepted philosophical base for the development of an environmental law system. As a nation we have no controlling ethical relationship with our natural environment; some persons wish to exploit, others to preserve, and most are in between. As a result, the basic framework of an environmental law system develops slowly, often with inconsistent rules on the same subject. And this development of law occurs in a political environment that is often acrimonious, for the basic values of some members of our society are challenged by the new rules dealing with the environment.

But as national attitudes change, so does the law. For example, wilderness in biblical references was the antithesis of a garden, and gardens were created by man overcoming nature. To the Puritans, wilderness was a hostile environment, a last refuge for sinners. Wilderness in 18th and 19th Century western literature had an evil connotation. Often the phrase "howling wilderness" was used. Today, we have a Wilderness Act, Wilderness Areas, a Wilderness Society, and wilderness conferences—all more or less in praise of wilderness. Other concepts relating to the environment have been subject to similar changes of attitude.

An example of these attitudes can be seen in what we today generally regard as a national treasure—the Grand Canyon. Yet the first western man to see the canyon, don Lopez Cardenas, did not even write of his journey. The first report, by Pedro de Castaneda, gives a vivid account of the difficulties of the descent into the canyon, but not a word of its beauty or grandure. Three hundred years later, US Army Lt. Joseph Ives explored the canyon. He concluded that the Colorado River at the canyon's bottom would be an economical avenue for the transport of supplies, but that the area was a profitless locality which would forever be unvisited and undisturbed.

Ten years later, John Wesley Powell, the first man to follow the river through the entire canyon, described it as the "Grand Canyon." In 1903 Theodore Roosevelt called the canyon a natural wonder unparalleled throughout the world, and sought to protect it. In the 1960s the US Bureau of Reclamation sought to dam the river in the National Park to provide electricity. The sale of this power would have produced profits which would have been used to subsidize agriculture in the Southwest. Thus the history of the Grand Canyon is an example of the changing attitudes of man to nature.

Nature was once thought to be the enemy against whom man struggled to survive. Later nature became the source of exploitive wealth. In recent time we have begun to appreciate the relationship of man with nature, and man's dependence on the rest of nature. This "newer" theory, which was not exactly unknown among some primitive groups, calls for man to utilize nature's bounty, but to exist so that the dynamic natural system survives.

The historical development of environmental law parallels the changing philosophical view of nature. Prior to the beginning of the 20th Century, the legal system was used to encourage the development and exploitation of our natural environment. The government encouraged canal building, and gave away millions of acres of public lands to those who would exploit them. Beginning in the late 19th Century, the concept of conservation became popular, and this resulted in legislation in the early part of the 20th Century to conserve and manage our natural resources.

But conservation means many things to many people, and shortly after the conservation movement began, it split into two camps or schools of thought—a development that to some extent continues today. John Muir (American naturalist, 1838-1914) led the protectionist or nonexploitation conservationists, while Gifford Pinchott (American forester and politician, 1865-1946) was the leader of the "careful extraction" or "wise-use" school of conservation.

Because the words commonly used to define conservation can be so easily distorted and misused, it is probably wiser for the seeker of truth to look for basic principles, rather than to cling to iron rules that may be easily turned around. The basic principle of conservation has been expressed in many hundred of ways, most long before conservation became popular. Some aspects of the ages-old nature religions were conservationist at heart; others caused vast ecological destruction in the worship of nature. For example, the Teotihuacan people of Mexico's Classic period destroyed the forests, very likely to make lime for stuccoing their huge "City of the Gods," contributing to the drop in the water table and the drying of their corn fields—all while worshipping various nature gods and goddesses, including the goddess of water. Such destruction of the environment has been exceedingly common wherever man has multiplied to the point where his actions could strongly affect the environment. The basic attitude, therefore that man is a part of nature and that his well being depends on the survival of the natural environment is not enough. To fully understand the natural environment we must depend upon science.

Conservation is a way of life. Needless to say, few people are willing to subscribe to the return to nature plea. But it should be made clear that living in harmony with nature need not require the giving up of many of the advances that a thousand years of applied science and technology have provided. Science has to do with discovering true facts and relationships, but technology cannot claim the authority of science. Science requires the exclusion of the human factor to discover truth, and is the antithesis of humanism (72).

As we begin the 21st Century, the overall status of US environmental problems and solutions should be assessed in recognition of the following scientific facts:

- No new pollutants have been discovered in 20 years.

- No new pollutant health hazards have been discovered in 15 years.

- There is no threat of mass wildlife or habitat destruction in the US.

- There is no threat of mass land, air or water pollution in the US.

- All US pollutants and natural resources are controlled under at least one jurisdictional level of government environmental law.

Chapter 1

The Movement

The environmental movement emerged in the early 20th Century as a collection of conservationists, naturalists and bird watchers. The movement grew slowly until government began recording natural resource impacts of human activity, such as over hunting, timber clear cuts and strip mining. In the last half of the 20th Century the movement exploded worldwide on government recognition of the potential health effects of environmental transgressions. Today, one is considered to be uncivilized if unconcerned about the environment. Today, environmental matters are a free-for-all of political pomposity.

T.R. to Toxics

It has been said that the original environmentalist was our 26th President, Teddy Roosevelt, who established the National Park Service in the early 1900s to set aside wilderness lands and conservation areas. During his 1901 to 1909 Republican presidency, Roosevelt designated 150 national forests, the first 51 federal bird sanctuaries, 5 national parks, the first 18 national monuments, the first 4 national game preserves and the first 21 land reclamation projects. He placed 230 million acres of land under federal protection (27). Teddy Roosevelt was the first to use the word "conservation" in describing US government policy. Roosevelt had come to the view that "the irresponsible use of natural resources is a fundamental problem which underlies almost every other problem of our national life" (13).

With two World Wars and The Depression intervening, it would take nearly sixty years to complete a federal government framework for conservation (13). Initially, the concern was smoke, sewage, and such. The clean air, clean water, and solid waste acts of the 1960s were still animated mainly by aesthetic insults or the appearance of being dirty, offensive or untidy. As with conservation, the principal objectives were still visible, tangible and use-oriented. Clean air, swimmable water, and the proper containment of solid waste dumps were emphasized. For example, New York City isn't the Grand Canyon, but the clarity of its air is still worth conserving, if only because a city can be a scenic asset. The Endangered Species Act, passed unanimously by the Senate in 1973, seemed to come from the same old conservationist perspective. Cougars, bears, buffalo, and wolves were to be protected. Teddy Roosevelt would surely have voted to protect these animals from extinction as eagerly as he would have hunted them when they were abundant (13). These laws extended the simple conservationist philosophy to its logical, practical and measurable limits.

US Government is formally committed to encouraging "productive and enjoyable harmony between man and his environment." This promise was made in the National Environmental Policy Act of 1969 (NEPA). Passage of the NEPA in 1969 was an attempt to create a new frame of reference for the consideration of all major activities by the Federal Government: a frame of reference that would include consideration of impacts to the environment. To do this was to attempt to internalize within each government agency processes which should require real thought concerning environmental impacts. NEPA was the culmination of a decade of previously unsuccessful Congressional attempts to define and put into practice a national environmental policy. The Resources and Conservation Act, proposed in 1959, called for the Executive branch to coordinate its scattered conservation efforts. The Ecological Resources and Surveys Bill of 1966 contained provisions designed to remedy the inadequate use of environmental data by federal agencies.

The original 1969 senate bill which was to become NEPA, Senate Bill 1075, was introduced by Senator Henry Jackson of Washington. It had a three-fold purpose: to establish national environmental policy, to authorize research concerning natural resources, and to establish a council of environmental advisors. As introduced, Senator Jackson's bill did not contain any operational procedure to assure implementation of a national environmental policy. Although the language called for "a national strategy for management of the human environment," it did not provide any specific procedures for review, coordination or control of continuing activities or decision-making. It was after the single-day hearing on S. 1075 before the Senate Interior and Insular Affairs Committee that the provision for an "action-forcing" or operational measure was added to the bill. This added section was number 102(2)(C), which required Federal agencies to prepare an "environmental impact statement" if any of their proposed actions might significantly impact the environment.

One purpose in enacting NEPA was to fulfill the need for an interdisciplinary approach to environmental management and decision-making in all branches and levels of the Federal Government. Previously, President Nixon had attempted to improve the Executive organization through his Executive Order of May 29, 1969, which created an interdepartmental, Cabinet-level Environmental Quality Council and a Citizen's Advisory Committee of Environmental Quality. NEPA provided new approaches for dealing with environmental problems on a preventative and anticipatory basis, and represents a break from the previous practices of dealing only with environmental crises and attempting to reclaim resources from past abuse.

NEPA did not create the kind of organization that was envisioned originally by its Congressional midwives. The concept of the Council on

Environmental Quality as an independent group of environmental wise-men, charged with viewing the national situation broadly and setting goals accordingly, has largely disappeared. The Council as created by the Act is an arm of the Office of the President, and as such speaks for the President. Its position as an independent voice is thus compromised by its relationship with the President: while the Council influences policy, policy influences the Council. Yet it is clearly undesirable to present political decisions resulting from less than disinterested wisdom. We had made great strides in our efforts to present to the President on a systematic basis essential environmental information. But a government-sponsored independent environmental watchdog has yet to be unleashed (72).

Even as the Federal Government completed the regulatory framework for traditional conservation, these laws also quietly launched the new era of environmentalism to control pollutants. To begin with, regulating pollutants of any kind requires a more elaborate and intrusive regulatory force than regulating the more human-scale parks, wilderness lands and wildlife. In 1970, President Nixon established a new cabinet-level agency under the US Department of the Interior, that would be the US Environmental Protection Agency (EPA). The EPA was established to take charge of all US pollution control and associated regulatory programs. More significantly thereafter, each of the new laws also included something quite new—an open-ended "toxics" provision, a general invitation for the EPA to monitor remote and invisible environments for hazardous pollutants and regulate them as needed for public health and habitat protection. In addition, though written with cougars and wilderness mainly in mind, the Endangered Species Act of 1973 (ESA) had been similarly expanded broadly enough to protect such exotics as the flower-loving fly, the fairy shrimp and other critters whose actual numbers in their natural habitats can never be counted. The ESA would soon be expanded to cover "habitat modifications" as well. For example, recently, the federal government designated the Swamp brush rabbit and Swamp wood rat as "endangered species"—the rabbit and rat only exist in a public park in Modesto, California (2). The US Fish and Wildlife Service (FWS), that determines species listings, may have turned itself into the de facto land use regulator. It appears the FWS first looks for a habitat it would like to save and then searches out an "endangered" species to justify invoking the draconian provisions of the ESA. Such practices of species protection law have become so routine that even some animal lovers have begun to question whether the ESA has been hijacked from its original purpose. In 1997, a National Wildlife Institute study found that in the ESA's 25-year history not a single listed species had fully recovered and been removed from the list, despite Federal efforts on its behalf (66).

A mere statutory afterthought in the 1960s, remote and invisible environmental pollutants get entire Federal Laws of their own a decade later (13). The Toxic Substances Control Act is implemented in 1976. Then the Superfund hazardous site cleanup program in 1980, followed by RCRA, CERCLA and SARA and other ubiquitous acronyms for government regulatory expansion to monitor and control every detectable impurity, which technology now allows us to identify at concentrations of less than one part per billion. For example, think of a football stadium filled to the very top with white marbles, with the ability to detect a single red marble among them, and then speculate with computer model projections that the single red marble at some concentration, at some distant, future time, is a hazardous pollutant that should be controlled by the Federal Government. You are reminded that the EPA began with a budget of $1 billion and 4,000 employees in 1970. Today, the EPA's budget is about $8 billion with 18,000 employees (4). This does not include the thousands of contract regulatory researchers operating under the EPA-administered grants whose millions of dollars will never be accounted for.

New Environmentalism

Somewhere between the Vietnam War and the Love Canal property contamination episode near Niagara Falls, the legal infrastructure of the "new
environmentalism" slips into place (13). Conservation isn't abandoned. It
is just overtaken politically, subsumed into something bigger. Bigger, par-
adoxically, because it concerns the very small, the remote, the invisible yet
detectable chemical agent whose environmental impact, quantifiable or
not, can become the fear of the moment for environmentalists and its
media exploitation with fertile opportunities for the cancerous expansion
of the government regulatory bureaucracy. Environmental regulations are
proposed and published every government business day in the Code of
Federal Regulations (CFR) at the rate of about 200 pages per day. In
1999, 24 volumes of environmental regulations, including 28,000 revi-
sion entries, were printed in 75,000 pages of the Federal Register (FR) by
the US Government Printing Office (8).

Environmentalism, and its environmentalist believers, didn't become a
potent public political movement until the 1960s and 70s in the US when
college campuses, brimming with idealistic baby boomers, were determined to make every new emotional twitch a political movement—a
cause for revolution. This is when political movements, valid or not,
became television news programming assets, and when anti-establishment
and counter cultural influences became media partners in a way that is
largely taken for granted today. Today, news and entertainment mass
media have become indistinguishable and readily exploitable as the propaganda machine in the game of ecopolitics (24). In today's ecopolitics,
environmentalists convene as media events. Politicians arrive to say "they
care" to placate the environmentalists for short-term media and electoral
rewards, enabling the irrational regulatory infrastructure to expand without end or accountability to measurable results. Environmental policy has
been all but separated from scientific evidence. Environmental policy is

now achieved through regulatory fiat to sate political activists who themselves can no longer be bothered with letting true science get in the way of what they want (42). Here is a tragic legacy of the 1960s and 70s cultural revolution; where the "rhetoric of virtue" and well-intentioned compassion-baiting dialogue make truth a negotiable commodity to advance one's public political ideology (64). This is a malevolent measure of the corrupt, dysfunctional state of US ecopolitics in the 21st Century.

There are three basic and sustaining, yet fallacious, public perceptions that the environmental movement nurtures and exploits, with man as associated with western democracies being the villain. These public perceptions—indeed misperceptions—persist because valid environmental analysis ultimately reduces to science, not feelings of compassion, guilt, fear, narcissism or political opportunism. The public and its media have little interest in, or understanding of, science, and are easily manipulated by anyone with the facility for emotive scientific jargon to promote controversy and insecurity among the citizenry.

First, there is the flawed public perception that **all** local environmental impacts, irrespective of their relative scale or ecology, are somehow cumulative to, and compounded with, **all** other known environmental impacts to result in a critical, irreversible regional or global environmental collapse. There simply is **no** empirical example of any such terminal collapse occurring or likely to occur as a result of human activities. The expanding ripples on 19th Century writer/educator/preservationist Henry David Thoreau's *Walden Pond* were contained succinctly within the pond water, without impact to air or land. Even if one of the three natural resources of land, air or water is impacted, its condition does not necessarily add to, or compound with, other impacted resources in any grand global cataclysm. Provocative speculation that **all** impacts to the natural resources of land, air and water result in significant, irreparable ecological impairments is simplistically and functionally flawed.

Second, there is the public misperception that **all** living ecological systems are inherently fragile and mysteriously complex. Living ecological systems are powerful and wondrous forces whose resilience and adaptability far exceed man's relatively brief tenure and influence on earth. Man has a tendency to conceitedly view all living systems in human consciousness; to anthropomorphize nature. Unlike man, nature acts only in the elegant efficiency of survival without ideology, morality, economics, politics, psychology, compassion or historical reflection. Nature is both intrinsically final and infallible. Ecological systems have evolved exquisite assimilative capacities, mutations, bioremediations, dispersions, redundancies, regenerative, recovery and energy management capabilities. Ecological systems are not inherently fragile. The contrary is true. The scientific record of life on earth demonstrates comprehensively and conclusively that ecological systems can and will sustain themselves infinitely in time, form and among all natural resources.

Third, there is the public perception (again, misperception) that business and industry enterprises have a vested sinister interest and covert mission to destroy and pollute natural resources without regard for ecology or human health. Business and industry are motivated by profit, not pollution. In the US, profit and pollution are ultimately incompatible goals for long term business and industry survival. Legitimate business and industry have adopted environmental compliance as part of their routine production and public relations objectives. US business and industry have demonstrated over the last 30 years that they will address environmental issues as long as regulatory controls are applied consistently among their competition, and as long as regulations are based in rational science with measurable positive results.

Environmentalists and their groups do not make their reputations or raise funds by making public pronouncements about the enormous progress in the US of solving environmental problems. What environmentalists don't want you to know is that almost every thing and activity comes under environmental regulation in the US and most pollution problems are solved or under active management.

Chapter 2

Environmentalists

The environmental movement has members whose daily lives involve a conscious self-reflection about human uses, conditions, consumptions, products, ambitions, ideology and survival as negative impacts upon nature. This self-reflective being, with rational to wildly irrational motivations, is an environmentalist.

Ethos and Id

Who are environmentalists? What do they believe? What do they want? Well, demographically, they're mostly middle to upper-middle class white, college-educated, agnostic or un-religious, ages 17-30, and seemingly well-intentioned people with **lots** of time on their hands. Most have never taken on the responsibility of parenthood or running a business. They generally rely on academic and government institutions for their wisdom and values, and prefer group identity over self-reliance. Environmentalists have a romantic fixation on simplicity and even passivity as rules for living in a complex and energetic western culture. Ironically, simplicity and passivity make life poorer, not greener (13).

Radical environmentalists are intolerant of growth, prosperity and free enterprise. Eco-fundamentalists rabidly resist measuring their goals against other critical concerns, like economics. They reflexively detest competition, capitalism, political diversity, corporate globalism and personal ambition beyond primitive, neolithic *hunter gatherer status* that ended 10,000 years ago for most of the human species. Radical environmentalists are activist ideologues, not environmental analysts, scientists or engineers, and at their fanatical extremes their interests conflict with immigrant and minority rights. This is because environmentalists are against housing developments that improve housing affordability, and are against global trade that provides economic growth and jobs abroad. For example, in 1999 the venerable Sierra Club, which was founded in 1892 and now has a half million members, considered a nationwide member referendum on whether the Club should take a position to limit US immigration due to its impacts of US population growth and consequent demands for space, natural resources and quality of life. This sort of *political mission creep* into *class warfare* has left such organizations trivialized and fractious.

Environmentalists are blindly resentful of the broad growth in economic opportunities that comes from free enterprise. Environmentalists see themselves as heroic (if not messianic) figures in a movement that is more socialistic than problem solving. This is where the environmental ideology can take on the trappings of a religious crusade. The movement transcends the need for truth to aid proselytization. Often, environmentalists' personal identification with eco-advocacy is a public attempt at redemption for a lack of virtue and worth in their individual personal lives—this is a peculiarly 1960s and 70s American cultural revolutionary notion.

Environmentalists can take many forms and degrees of activism, ranging from the civic gadfly who is a fixture at your city council meetings, to lifestyle gurus, to tree-hugging naturalists and NIMBY ("Not In My Backyard") whiners; to the extreme anti-property rights, anti-social, anti-technology, fanatics who subscribe to the beliefs of the infamous *unibomber*, Ted Kazynsky. This is a "*Luddite complex,*" historically associated with a group of early 19th Century English workmen who destroyed laborsaving machinery as a protest against technological advancement (37).

Environmentalists have gained an American pop-cultural identity. There is a humorous side to the cliché, stereotypical, eco-creature we have come to identify as the *environmentalist*. The following comical behaviors may be observed in environmentalists:

They might be an *environmentalist* if...

— their idea of luxury accommodations includes indoor plumbing and a free water purification tablet.

— their desk calendar has the World Trade Organization meeting schedule highlighted through the year 2050.

— they flush their toilet once a week whether it needs it or not.

— they personally rotate their tires every 10 days.

— they've been using the same grocery shopping bag since the Carter Administration.

— their kitchen card file has a recipe section under the heading "Roadkill."

— their wedding included Pete Seeger songs and vows written by Ed Begley, Jr.

— they make their own ethanol fuel from stale Doritos.

— they describe Ralph Nader as a passionate, charismatic and spiritual national figure.

— they're a bored successful actor who lives and works in a fantasy world, and needs a huge tax deduction.

- they socially identify with the extinct Neanderthal hunter-gatherer class.

- their political activism promotes passivism without a spiritual component.

- their greatest political achievement was painting "Prosperity Sucks" picket signs for a Green Party rally.

- they operate a flatulence recovery unit for their neighborhood to reduce greenhouse gases and global warming.

- they're unable to identify areas of land that should be designated as "suitable human habitat."

Prime Movers

An original scholar in the theories of man's conflict with the sustainability of natural resources (the environment) was Thomas R. Malthus (1766-1834). Malthus was an English clergyman and economist (36). Malthus published an anonymous pamphlet in 1798 promoting the theory that human population increases geometrically, while food supplies can only increase arithmetically. Therefore, sooner or later, the growing gap between food supply and food demand must end in war, famine and general human misery. Malthus simply argued that when mankind reaches the limits of nature—when it had farmed all the farmable land—mankind would starve. With the technology of 1798, Malthusian theory was both obvious and true in postulating that the ascent of man causes the collapse of everything else and that, in turn, destroys man too (13). Now, 200 years later, clearly Malthus was in error because he grossly underestimated man's ever-evolving ingenuity. Human resourcefulness was left out of the equation.

Born and raised on a farm in Springdale, Pennsylvania, Rachel Carson wrote a seminal book on the overuse of US pesticides in agriculture titled *Silent Spring* in 1962. DDT and its derivative chlorinated hydrocarbons were agricultural pesticides developed in the late 1930s. DDT was widely used in farming to improve crop yields, and was found to control almost any type of insect. These chlorinated hydrocarbons are quite resilient in the environment, having decomposition half-lifes of 10 to 15 years (48). Though not an ecologist, Carson's book implicates reckless worldwide pesticide use in the unraveling of ecosystem food chains in very personal, almost romantic language. Carson predicted that pesticide use would eventually contaminate world drinking water supplies. DDT use in the US was banned by Federal Law in 1972 (36). Carson's prediction, fortunately, was wrong.

A more consequential environmental scholar, probably because he wrote during the 1960s and 70s era when college campus political movements *flowered*, is Stanford University biologist Paul Ehrlich. In 1968, Ehrlich wrote a bestseller titled *The Population Bomb*. Ehrlich's book embraces and extends the Malthusian theory toward the irrational. Ehrlich's 1968 doomsday vision stated, "The battle to feed all of humanity is over. In the 1970s and 1980s, hundreds of millions of people will starve to death in spite of any crash programs embarked upon now." Time has proven Ehrlich to be flamboyantly and profoundly wrong. He admonished that nature will take its revenge against mankind's abuses. Further, Ehrlich attributes the AIDS epidemic to the "deterioration of the epidemiological environment which is quite directly related to population size as well as to poverty and environmental deterioration" (13). Ehrlich appears to have stumbled from his basic life science expertise into elitist psycho-socio-babble, to his professional discredit.

US Senator, Vice President and presidential aspirant Al Gore authored a 400-page tome on everything environmental titled *Earth in the Balance: Ecology and the Human Spirit*, which became a "national bestseller" (68).

Gore, in writing the book, said he was willing to risk his entire political career on the issue of *the environment*. His original 1990 title for the book was "World War III." Gore's emphasis was that to attain his version of global environmental rectitude would require the commitment and sacrifice of world war (69). *Earth in the Balance* rambles maniacally among premises of over population, nature's spirit, technophobia, consumptionism, species extinction and global warming in an attempt to circumscribe an "environmental holocaust without precedent" and to position Gore as the political leader whose insight will save the planet. Aside from his *Luddite* leanings, Gore exposes himself as either profoundly confused, or cynically manipulative, about the meaning of "technology." He writes "[G]overnment, as a tool used to achieve social and political organization, may be considered a technology, and in that sense self-government is one of the most sophisticated technologies ever created." His further abstractions equate technology with "spoken language," and even "the human body." Gore also elaborates on how technology is not necessarily science. One should always be suspicious when politicians begin to bend the meaning of words. For clarity, please observe that Webster's Dictionary defines technology as "applied science" (37). Gore also calls for science and religion to be "reunited in the service of the environment."

Gore, in keeping with the world war analogy of his environmental crusade, promotes vast government programs such as a "Strategic Environmental Initiative" and "Global Marshall Plan." His Global Marshall Plan would act to stabilize world population, develop environmentally appropriate technology, measure environmental impacts in economic terms, develop international environmental regulatory programs and develop a global environmental education program. His Strategic Environmental Initiative was named to imply an environmental equivalent of the "Strategic Defense Initiative" (SDI), the crash program to develop a series of technological breakthroughs focusing on a common military objective, which Gore opposed as senator. Gore's Strategic

Environmental Initiative would be a global "program that would discourage and phase out older, inappropriate technologies and at the same time develop and disseminate a new generation of sophisticated and environmentally benign substitutes." This, from a politician who clearly does not understand the definition of the word "technology."

Gore further writes to introduce the divisive environmental concepts of "environmental justice" and "sustainable development" for environmentalist promotions. These latent environmentalist concepts are attempts to leverage social issues of immigration, class warfare, racism and big business bashing for pure political patronage. Gore climaxes his pretentious and partisan policy masturbations with a renewed self-assurance in big government dominance over personal ambition and free enterprise.

A more recent example of the endless desire to identify with *things environmental* is the pseudo-scientific, psycho-social, co-optation called *ecopsychology*. Every political movement has its psychological dimension. This is no less true for the environmental movement. Persuading people to alter their behavior always involves probing motivations and debating values. Political activism begins with asking what makes people tick. What does the public want, fear and care about? How do we get and hold the public interest? How much can people take; what are their priorities? Have activists overloaded the populace with anxiety and guilt? Ecopsychology endeavors to address the problem of effective communication with the general public in order to meet the demands of the "environmental revolution." Ecopsychology claims to "redefine sanity within an environmental context," to re-examine the human psyche as an integral part of "the web of nature" or ecosystem. Ecopsychology presumes to "bring together the sensitivity of psycho-therapists, the expertise of ecologists and the ethical energy of environmental activists" (46). Here is a perfect example of the abstraction of a physical science discipline (i.e., environmental science) with a soft social science discipline (i.e., psychology). This is sophistry, not science, and certainly no basis for making prudent management decisions about our natural resources.

Darker Green

Most environmentalists are volunteers, but the environmental movement supports professional activists in issues where environmentalists become government or tax-exempt activist group staff and administrators. The presidents of the Environmental Defense Fund, Sierra Club and The Wilderness Society make about $125,000 per year. The president of the Natural Resources Defense Council makes about $150,000 per year. The president of the World Wildlife Fund makes about $200,000 per year. The head of the National Wildlife Federation makes approximately $250,000 per year (21). Given these levels of income, it appears that some of the *prophets of doom* regarding the environment are profiting from doom.

The following are profiles of the US's largest, most visible environmental groups (63):

- **Greenpeace**, *Amsterdam, Holland*
  Known for radical actions, such as sailing into nuclear-test zones, to publicize concerns on world-wide issues like global warming, food toxins and whaling.
  Membership—350,000.

- **Earth First**, *No headquarters disclosed.*
  Loosely organized group that engages in direct, and often civilly disobedient, actions such as forming human chains to block logging roads.
  Membership—Has no official roster.

- **Friends of the Earth**, *Washington, D.C.*
  Progressive group with broad international focus on issues like global warming; works through political process and lobbies for policy change.
  Membership—Not disclosed.

- **Sierra Club**, *San Francisco, CA*
  Mainstream, but increasingly extreme, organization with vast membership and emphasis on US; uses political process, getting members to petition lawmakers; sues.
  Membership—500,000.

- **Natural Resources Defense Council**, *New York, NY*
  Developed over time by many lawyers for various environmental groups; engages in litigation to force environmental protections.
  Membership—170,000.

- **National Audubon Society**, *New York, NY*
  Actively campaigns for protection of bird habitat across the US; maintains chapters around the country and is active in local politics.
  Membership—550,000.

- **Wilderness Society**, *Washington, D.C.*
  Primarily focused on land protection, such as in the vast federal holdings in the western US; lobbies Congress and other federal agencies.
  Membership—200,000.

- **National Wildlife Federation**, *Vienna, VA*
  Originated as collection of hunting and fishing groups; has strong nature focus; issues include clean air; lobbies for legislative change; works with government agencies.
  Membership—4 million.

- **World Wildlife Fund**, *Washington, D.C.*
  Focuses on conservation projects such as protecting wildlife; advocates legislative change.
  Membership—1 million.

- **The Nature Conservancy**, *Arlington, VA*
  Chiefly interested in protecting threatened habitat; buys threatened land from private owners and resells to federal government.
  Membership—900,000.

Don't be fooled into thinking that most environmentalists are green-robed, poverty-vowed, sandal-shod, do-gooders who wander the woods to prostrate themselves before the evil bulldozer for the sake of birds and bunny rabbits. The movement has become politically sophisticated and opportunistic. The environmental movement no longer has the time for true science to identify true environmental risks. It is interesting to note that major urban daily newspaper classified advertisements and certain websites routinely solicit workers under the help-wanted advertisement heading "Activist." The *Los Angeles Times* classifieds ran extensive ads soliciting activists on a range of environmental topics prior to, and following, the August 2000 Democratic National Convention in Los Angeles. The nation's largest metropolitan daily paper, under "Jobs Offered", Section "8500", between "Accounting" and "Admin. Asst." advertised "Activists", "no exp. O.K." for the Sierra Club and other groups. Up to $700 per week was offered to convert anyone, without experience, into a mercenary environmental activist. There is more *green* than meets the eye in the commercial side of environmentalism.

As nationally or internationally organized groups, environmentalists can become radical, extreme, anarchist or nihilist entities such as Greenpeace or Earth First with truly eco-terroristic tendencies. While Greenpeace is widely known for its eco-stunts of installing human chains and large protest signs on ships, trees, skyscrapers and bridges that jeopardize lives and property, something even more disturbing called the Earth Liberation Front has in recent years taken responsibility for fire bombings of new housing development construction sites in the Midwest. You are reminded that most of these environmental groups operate under your tax subsidy as non-, or not-for-, profit organizations.

The fuzzy regulatory area of non-, or not-for-, profit tax-exempt organizations deserves scrutiny as regards the environmental movement and its groups. Most of the well-known environmental groups operate as tax-exempt organizations under US Internal Revenue Service (IRS) Code Sections 501(c)(3) and 527. These groups include the Sierra Club, National Resources Defense Council, National Wildlife Federation and others with IRS tax-exempt status that are effectively subsidized by all US taxpayers. The IRS tax-exempt status infers that these groups are taxpayer subsidized because they act in the *public interest* to enhance and protect *public welfare* and they therefore deserve the *public trust*.

General categories of IRS tax-exempt organizations include those organized and operated exclusively for one or more of the following purposes: religious, charitable, scientific, testing for public safety, literary, educational, protection of children or animals, or amateur sports promotion (47). Few of the nationally-dominant tax-exempt environmental organizations can be seen to strictly qualify under one or more of the foregoing IRS categories. Many of the nationally-prominent tax-exempt environmental organizations frequently jeopardize their tax-exempt status under the IRS "propaganda" prohibition at IRS Code Title 26, Subtitle A, Chapter 1, Subchapter F, Part I, Section 501(c)(3) that states, "…(organization is exempted provided) no substantial part of the (organization's) activities…is carrying on propaganda or otherwise attempting to influence legislation or intervene in any political campaign on behalf of any candidate for public office." Propaganda is defined as ideas, facts or allegations spread deliberately to further a cause or to damage an opposing cause (37). Recently, environmental organizations, including the Sierra Club, have been cautioned by the IRS for lobbying and propaganda activities that exceed the "no substantial part of activities" tax exempt status standard under IRS Code Sections 501 and 527. In the summer of 2000, the IRS began requiring Section 527 tax-exempt organizations to report "soft money" flowing through political campaigns. Such organizations must report a variety of data including their contributions, their expenditures and their purposes. In the first two months of the new Section 527 reporting requirement, over 8,000 "political action committees" (PACs), including environmental groups, filed reports (50).

Over the last 30 years, environmentalists have done a good job of frightening and shaking the general populace into first the hysteria of "the sky is falling," then more recently into resentment over the exaggerated daily claims of environmental apocalypse, until today where warning-battered rational people are barely listening to environmentalists. Note the lack of *the environment* as a serious presidential campaign issue in the 2000 elections. Even Ralph Nader did not include the word *environment* in his Green Party 2000 presidential nomination acceptance speech on June 25, 2000.

After years of being bombarded by ever more dire ecological prophecies, of which **none** has materialized, Americans have grown more and more skeptical of environmentalist predictions and protests. The alarmist theories that environmentalist made fashionable during debates over acid rain, toxic groundwater, nuclear winter, over-population and species extinction, have backfired. Americans have now adopted a "selective deafness" as a first line of defense against wild claims of environmental disaster (22).

Ecopsychologist Theodore Roszak noted in his 1993 book, *The Voice of the Earth*, that the environmental movement might have overutilized shame-and-blame tactics in its approach to the public. And, that the public may become particularly vulnerable to right-wing conservative attempts to instigate a "green backlash." Roszak theorizes that the green backlash may provide people an opportunity to avoid feelings of guilt and helplessness, and to attack environmentalists who make them feel that way (46). Roszak's bizarre musings reveal the perverse and cynical environmentalist's manipulative devices of promoting public guilt and helplessness (*victimization*).

As the focus of environmentalists moves from once-immediate dangers now under control, to more abstract matters of aesthetics or "sustainable living," latent economic class conflicts are beginning to erupt. Pollution abatement often imposes highly regressive costs according to socio-economic class. For example, in the early 1990s, the cost burdens of Southern California's aggressive air quality management plans were estimated to have a three times greater impact upon the region's poorest households than on the wealthiest. Environmentalists dismiss such economic irritants by arguing that a better environment helps everyone. The proverbial ecological crisis notwithstanding, the adverse health consequences of reduced economic opportunities for the poor vastly overwhelm any environmental benefits they may enjoy from, say, marginally cleaner air quality. US air pollution reduces average life expectancy by approximately 30 days. Poverty strips away 10 years in life expectancy (1).

The experience of 30 years of environmental controls in the US is testament to conserved natural resources and solved pollution problems, and conclusively demonstrates that growth-oriented economies (i.e., free-market democracies) actually do a better job of managing natural resources than a society run on the myopic principles and utopian-directed theories of environmentalist dogma (13).

Chapter 3

Truth Is Secondary

Politics and truth seldom occupy the same space anymore. When the environmental movement found a political base, it began to leave the truth and scientific rationale for natural resource management policy behind as too cumbersome. In the climate of ecopolitics, the US environmental movement has lost its way and legitimacy in the 21st Century. Environmentalism has become just another partisan means to a political end.

Scares, Lies and Red Tape

In order to glean some truth from any activist representations, one must start from the cautious position that the first casualty of activism is the truth—truth is secondary. Activists, including environmentalists, rhetoric is tactically reliant upon exaggerated dangers and inflammatory word use to erect counterfeit arguments and promote bumper sticker platitudes as moral authority for social engineering. Typically, the *zero sum* false dilemma proposed is that of "it's only man **or** nature" that will survive the current environmental disaster (23). Anyone with only a rudimentary knowledge of the living systems of Planet Earth, that include both man and nature, understands that nature, with or without man, shall ultimately prevail just as it did for over 3 billion years prior to man's unceremonious and primitive appearance on earth about 3 million years ago.

The founding beliefs of environmentalists are that of scarcity, of limits to growth and therefore, "the sky is falling," "the end is near," "catastrophe is just around the corner" fear-mongering proclamations. Their basic and yet transparent proof of the collapse of the earth's ecosystem is blamed mainly on the American democratic way of life and its prosperity where actually only 5% of the world's population is involved. Their *guilt game* carries the further shame that Americans consume 28% of the worlds natural gas, 23% of the solid fuels, 20% of the coal, 23% of the crude oil, 42% of the gasoline, 26% of electricity, and 10 to 30% of its copper, aluminum and zinc, and drive more and larger vehicles more miles than any other nation (31). One can only surmise that the operative environmentalist theory is that "the high quality of American Life **must** somehow be a threat to the Planet Earth."

Where is the truth? Recent scholars in the true environmental sciences have coined the term *trans-science* to describe the study of phenomena that are too large, too diffuse, too rare, too distant or too long term to be resolved by reliable scientific methods. Exploiting trans-science, environmentalists release so-called "studies" that bypass the medical and scientific journals and peer review, and go straight to sympathetic, issue-hungry and largely gullible journalists (33). The due diligence of true scientific cause-and-effect findings is neglected.

Of course, without each of us applying some critical examination and context to these waves of scary news, the complex, remote and invisible threat **will** make you uneasy as it is designed to do. It is also designed to create civic anxiety, demonize business and industry, and thus promote a dependency on government to solve a problem that may not exist in the priorities of our lives. Consensus in legitimate science has **always** led to rational truth. Today, the radical environmental movement is more alchemy than altruism. Today, the media is more concerned with the controversy that attaches to environmentalism rather than truth telling about the state of the environment.

American courtrooms and government regulatory agencies have been overwhelmed by health scares linked to environmental issues—pesticides, ozone depletion, electromagnetic waves from cell phones, etc. For example, government-funded environmental scientists recently were found to have systematically distorted data to support the theory that the invisible electromagnetic fields near power lines cause cancer. Is their distortion a surprise? Too often, government environmental researchers operate as a self-selected, insular, academic group with an inherent bias toward identifying industrial environmental pollutants, particularly from American industry, as imminent public health risks. Often, this leads them to become personally and emotionally wedded to their own theories, and scientific objectivity is lost along with the truth (34). In the example of electromagnetic fields causing cancer, it is estimated that lawsuit compensation awards, needless mitigations and unnecessary research costs amount to $25 billion for this environmental health scare (41).

Today, a successful government regulatory scientist may find possible environmental problems, and then publicize them as probable public health risks to get the attention of legislators for follow-up work and renewed funding (19). Often the incentives in government are to save, rather than solve the problem, and thereby save bureaucratic power and it's vast taxpayer-supported *civil service* employment opportunities (17). You are reminded that the US Government, with the world's most extravagant civil service employee benefits and security, is by far the largest employer in the US. The US governments' employee labor unions (local, state and federal) account for 37% of all labor union membership, one of the largest money contributor and voting blocks for the political left (26).

EPA Missteps

Today the public, by and large, seems satisfied with the successes of environmental control. But the EPA is not satisfied with environmental controls. If the pollution problem is largely solved, what is the EPA to do now to keep itself alive? The logical answer, and one that all bureaucracies, public and private gravitate toward, is to invent new problems. In the EPA's case, the best way to do that is to continue to expand restrictive pollution standards toward the irrational, and to leverage the fear mongering of environmental dangers in an ongoing marketing campaign for government services. But the agency, which exercises broad discretionary powers over land, air and water resources and industry, as a result of the open-ended regulatory mandate it received 30 years ago from Congress, finally ran up against opposition in August 1999. A Washington panel of federal judges ruled, on behalf of a number of complainants including several states, that the EPA's latest standards for control of urban ozone and microscopic particulate air pollutants were the product of a selective use of science and showed little evidence of cost-benefit criteria having been applied. The Court concluded that, the "EPA offered no intelligible principle by which to identify a regulatory stopping point" (12). The EPA could offer no adequate justification for drawing "the requisite to protect public health" line for ozone and particulate pollutants. The EPA has appealed the case to the US Supreme Court, where the court will have to determine if the EPA exceeded its regulatory mandate by acting legislatively in violation of the "nondelegation doctrine" (43).

The EPA administers the Safe Drinking Water Act (SDWA), where so-called "Maximum Contaminant Level Goals" (MCLG) for public drinking water are established. The 1996 amendments to the SDWA revised the procedure to be used by the EPA in setting MCLGs to assure that the EPA uses "the best available scientific information." In 1994, the EPA proposed a "zero" MCLG for the chemical chloroform in drinking water, based on data that chloroform causes cancer. The Chlorine Chemistry Council and others petitioned the Federal Court of Appeals to review and repeal the setting of an MCLG of zero for chloroform, asserting that the EPA violated "the best available scientific evidence" requirement. The Court held that the EPA's adopting an MCLG of zero for chloroform was arbitrary and capricious, beyond statutory authority and void. Basically, the Court agreed that the EPA violated the statutory mandate to use best available scientific evidence. The Court said that the EPA can not reject the "best available" evidence simply because of the possibility of contradiction in the future by evidence unavailable at the time of action—a possibility that would always exist in any government regulatory action (44).

The EPA has issued rules under its broad discretionary mandate of the Clean Water Act (CWA) to require more stringent state action in improving the water quality of "non-point sources" or rainfall-induced runoff from urban, commercial, industrial and agricultural land uses. Here, certain "Total Maximum Daily Load" (TMDL) limits for pollutants in runoff were deemed by the EPA to be too weak even after Congress had enacted legislation to prevent the EPA from tightening the limits. The EPA nevertheless issued tougher water runoff rules that were described by congress members as "no more than a political power grab" with "no sound scientific reason for doing any of these things." The US Chamber of Commerce called the EPA's action an "underhanded, last-minute effort to thwart the will of Congress." The congressional and business opposition to the tougher water runoff rules centered on arguments that the EPA's precipitous rulemaking "usurps state authority and responsibilities, and imposes vast costs not justified by the potential results (benefits)" (53).

In early 2000, Congress held hearings concerning the use of "bullying tactics" by the EPA against its critics who are private citizens. A dairy farmer testified that she received threatening letters from the EPA water quality scientists after she spoke out publicly against the use of treated sewage sludge as fertilizer. The dairy farmer stated that she perceived the EPA regulatory scientists' letters to be a threat of more frequent, punitive Federal Government inspections of her farm unless she ceased her protests (54).

The EPA staff members were recently prosecuted for fraud in falsifying applications for a Tribal (American Indian) Partners Program for water pollution controls. In another instance, the Department of the Interior and FBI have reported that computer hackers have compromised the EPA computers since 1992, shutting down the EPA's web site. The US General Accounting Office (GAO) reported that hackers were able to tamper with data, browse sensitive information, and even attack other government agencies using the EPA computer systems. The EPA's ability to detect computer hackers was reported to be so flawed that it failed to detect the government's own security experts hacking test efforts (51). The GAO concluded that the computer operating systems and the agency-wide computer network that support most of the EPA's mission-related and financial operations are riddled with security weaknesses, stating the "EPA cannot guarantee the protection of sensitive business and financial data kept on its larger systems or supported by the agency-wide network" (54).

Perhaps the most egregious misstep in the EPA's reflex to regulate concerns the miscoordination among its land, water and air pollution programs. In 1984, the EPA's studies estimated that there were two million underground storage tanks (USTs) used for storing bulk supplies of petroleum products (mostly gasoline service stations) and certain EPA-listed hazardous substances in the US; a typical example would be your local gas station. Research determined that most of the USTs were of corrodible steel construction and that most of such USTs would leak their hazardous contents into surrounding soils or water resources after 20 years of burial. The EPA promptly established regulatory programs to prevent, detect and clean up leaking USTs (LUSTs). A nationwide compliance window of 10 years was ordered in 1988 to fully clean up the nation's LUSTs by 1998. You probably observed this program in action when your local gas station appeared one day to be closed and fenced, with piles of dirt as evidence of LUST removals.

Apparently, without regard to the foregoing estimates of LUST contaminations of soil and water resources, the EPA-administered Clean Air Act, and more specifically, the 1990 amendments thereto, required many large US city gasoline suppliers to provide reformulated gasoline (RFG) to reduce air pollution from urban car exhausts. One of the active ingredients in RFG is methyl tertiary-butyl ether (MTBE), an EPA-listed hazardous chemical compound because it is toxic and causes cancer. While the burning of MTBE-containing RFG in your car reduces the air pollutant carbon monoxide in your car's exhaust and atmospheric ozone, the MTBE-containing RFG is a hazardous substance when released to land, water or air resources. To meet the RFG demand for MTBE, 27 US companies began producing up to 10 billion pounds per year of the hazardous compound MTBE.

Pre-1998 LUSTs in cities using 1990-required MTBE-containing RFG have been discovered to have leaked their RFG into urban soils and water resources, contaminating public drinking water supplies with the toxic and carcinogenic MTBE. Thousands of drinking water wells and reservoirs from California, Maine and dozens of northeastern states tested positive for the EPA-required toxic and cancer-causing MTBE. Later, researchers found that MTBE's particular persistence and mobility in water, even groundwater, is more likely to contaminate public well and reservoir drinking water supplies than all other hazardous compounds in automobile gasolines. The City of Santa Monica, California has sued numerous oil companies for supplying the EPA-required MTBE-containing RFG that has contaminated the city's drinking water supply. The city's claim against the oil companies totals approximately $200 million in damages. Based upon these and other findings, the EPA began yet further rulemaking to eliminate or limit the use of MTBE as a gasoline additive in March 2000 (55). This lapse in the EPA pollution program coordination, and its penchant for impulsive regulatory action, has exposed tens of thousands of public water supply consumers to the toxic and cancer causing additive MTBE during the decade of the 1990s.

The EPA, established by President Richard M. Nixon in 1970, is the first, largest, most costly and far reaching national government environmental regulatory agency in the world. The EPA should be credited for developing and enforcing regulatory systems and fostering pollution control technology innovations that have in most cases benefited US ecology and human health when based upon firm scientific evidence.

As the nation's land, water and air continue to become cleaner, the EPA and other environmental regulatory agencies face reduced taxpayer-supported budgets and maybe even bureaucratic obsolescence. The problem is that "bureaucratic obsolescence" is descriptively oxymoronic; American government bureaucracies seem to have no proven path to extinction. Such agencies will therefore continually tighten regulatory standards so that reasonable compliance is never achieved. Today, there is an unholy alliance of environmentalists, media and regulatory bureaucrats that conjures up environmental evils, dresses them up as science in a system that is perfectly evolved to fund and grow the government establishment. There are legions of academics and regulatory scientists whose occupation it is to invade, critique, punish, and ultimately dictate your lifestyle.

Enviro-Mythmaking

There has been a profusion of books on *the environment* over the last 30 years. Unlike this book, they predominately cover "the sky is falling" mantra, and a call to guilt-driven personal sacrifice and recruitment to organize in the revolutionary cause of environmentalism. An interesting benchmark of environmental activism is the hysterionic doomsday propaganda that filled books released to coincide with the twentieth anniversary of Earth Day in April 1990. This date is a generation removed from the original Earth Day established by Republican President Nixon in 1970. The following referenced excerpts are examples of the hyperbolic (even psychic-like) claims made by authors of the environmental books that sought to capitalize on the book sales genre of Earth Day 1990. The *Earth in the Balance* book analyzed in preceding CHAPTER 2 should be considered in a class with the following books.

"The Exxon Valdez. Alar-treated apples. Love Canal. Bhopal. Three Mile Island. Chernobyl. The list of individual examples of environmental degradation could fill this book. But none of these calamities should be thought of as isolated incidents. Cumulatively, they add up to the worst state of environmental affairs since the dawn of civilization. And they've contributed to a fraternity of staggering problems whose scientific jargon has become the common language for consumers as well as scientists: global warming and the deterioration of the ozone layer, acid rain and urban smog, deforestation, toxic waste, garbage overload, water pollution." (*Save Our Planet*, Dell Trade Paperback, 1990.)

"Clean water is our most precious resource. But as much as a fourth of the world's reliable water supply could be rendered unsafe for use by the year 2000." (*Save Our Planet*, Dell Trade Paperback, 1990.)

"The Amazon basin, 3.1 million square miles of it, has the world's largest and most biologically diverse tropical forest. It contains anywhere from one-tenth to one-half of the planet's plant, insect, and animal species, depending on who is making the estimate. Edward Wilson of Harvard University calculates that forest destruction worldwide causes the extinction of about 10,000 species every year. In addition, one-quarter of all our medicinal drugs are derived from tropical forest plants. These include medicines that treat heart disease and childhood leukemia. We will never know how many other valuable drugs have been lost by forest destruction." (*Our Earth, Ourselves,* Bantam Books, 1990.)

"Fifty acres of rainforest are destroyed each minute. That's almost 27 million acres a year, an area equal in size to the state of Pennsylvania. At no time in history has the rate of deforestation approached what we are seeing as we enter the 1990s. Two-fifths of the world's original rainforest cover has been decimated, mostly in the last fifty years." (*Save Our Plant,* Dell Trade Paperback, 1990.)

"The Nature Conservancy says extinctions are accelerating worldwide. Our planet is now losing up to three species per day. That figure is predicted to be three species per hour in scarcely a decade. By the year 2000, 20% of all Earth's species could be lost forever." (*50 Simple Things You Can Do to Save the Earth,* Earthworks Press, 1989.)

"Eighty percent of America's solid waste is being dumped into 6,000 landfills, spread across every state in the country. But that option is shrinking fast. In the past five years, 3,000 dumps have been closed, and by 1993, some 2,000 more will be jammed to capacity and closed. In just four years, Chicago's landfills will be full; dumps in Los Angeles should reach capacity by 1995." (*Save Our Planet,* Dell Trade Paperback, 1990.)

"The Environmental Protection Agency estimates that in the next five to ten years more than twenty-seven states and half of the country's cities will run out of landfill space. Major cities including New York and Los Angeles will exhaust their landfill space in just a few years." (*Earth Right,* Prima Publishing, 1990.)

"When the federal Superfund Law was enacted in 1980, many people hoped that the federal cleanup program could be a short-term, one-time effort. It now appears that the task of cleaning up hazardous waste sites will haunt us well into the twenty-first century. Not only is it taking longer to clean up sites, but new sites continue to be discovered. Over 300,000 locations now contain hazardous substances, and the number is growing at a pace that far outstrips the rate of cleanup." (*Save Our Planet,* Dell Trade Paperback, 1990.)

"Some sixty US counties, including much of the urban Midwest and East, violate minimal air quality standards, spewing more pollution into the air than is legally permitted under the federal Clean Air Act. The American Lung Association believes that about 115 million Americans are being exposed to treacherous air pollution levels. The American Academy of Pediatrics believes that as many as 28 million children have been put at risk because the air is too dirty to breathe safely." (*Save Our Planet*, Dell Trade Paperback, 1990.)

"Even if the entire world were to stop using CFCs and halons immediately, destruction of the ozone layer would go on for decades.... The destruction of the ozone layer over Antarctica from CFCs and halons used today will continue into the twenty-first century. The Natural Resources Defense Council estimates that even with an immediate total ban on ozone-depleting chemicals, recovery of the ozone layer will take more than a century." (*Earth Right*, Prima Publishing, 1990.)

"In 1985, a 'hole' was found eating its way across the sky above Antarctica. It is now believed that this hole is as deep as Mount Everest is tall and as wide as the United States. Most scientific researchers are convinced that global CFC emissions must be reduced substantially, if not completely, to avoid a catastrophic depletion of the stratospheric ozone layer." (*Save Our Planet*, Dell Trade Paperback, 1990.)

Note the over-hyped language use such as: "calamities," "worst state of environmental affairs since the dawn of civilization," "staggering problems," "unsafe," "forest destruction," "destroyed each minute," "decimated," "species could be lost forever," "option is shrinking fast," "jammed to capacity," "will run out of landfill space," "waste sites will haunt us," "spewing more pollution," "treacherous air pollution," "28 million children put at risk," "ozone hole found eating its way across the sky," "catastrophic depletion." This dramatic language sounds more like Hollywood horror movie advertising tag lines than enlightenment about managing our natural resources. Needless to say, **none** of the preceding eco-horrors ever occurred or is now predicted to occur.

CHAPTER 6 herein, *10 ENVIRONMENTAL TRUTHS*, presents recent fully-referenced scientific findings that dispel the foregoing and other popular environmental myths, exaggerations and erroneous predictions.

Psychiatrists, psychologists and sociologists have long postulated that symptoms of irrational fear and anxiety increase when political and economic systems are most stable. This is why US institutionalized enviro-mythmaking has proved to be quite powerful, persuasive and pernicious. The involved institutions are mass media news sources such as TV, radio, the internet, magazines and newspapers. The basic, raw material for mass media is controversy, to feed the activity of fear mongering. University of Southern California Professor of Sociology, Barry Glassner, wrote a seminal and best-selling book on fear mongering in 1999 (57). Glassner debunks (among others) the 1990s enviro-myth of the dangers of US schools containing asbestos. The EPA estimated that one-third of the nation's schools contained asbestos insulation that when inhaled over long periods of time can cause lung cancer. That school kids would be exposed to the asbestos health risk became a public outrage. US schools spent over $10 billion to remove school asbestos even though its removal posed a greater health hazard risk than allowing the asbestos to remain installed and immobile. In this case, media engaged relentlessly in the school asbestos health scare as fear mongering; as they have with other health scares such as AIDS, Dow Corning silicon breast implants, Gulf War Syndrome, road rage and others. Why do fear mongering media campaigns take hold? Why do media and their audiences get drawn to one hazard over another?

Fear mongering motivates people to 1) correct a moral offense or 2) to criticize a disliked group or institution (58). Health hazards at any degree of injury or prevalence in the population are deemed to be morally unacceptable in the US, whether merely perceived or real in scientific medical terms. Witness the school asbestos example above as a motivating environmental health hazard. Some of this motivation arises out of a collective cultural narcissism or sense of entitlement; because real or not, health risks are innately personal and have resonance with our basic survival instincts.

The second fear mongering motivation to criticize or discredit a disliked group or institution came into full influence in America during the 1960s and 70s. Then til now, demonizing groups or institutions has become the activists' sport and even occupation to exploit environmentalist distrust of American corporate free enterprise—particularly "big business." Given the high threshold for motivating moral outrage and the seeking of personal redemption via public political protest that have characterized the political left, they are not likely to be motivated by an aforementioned moral offense in response to environmental issues. Rather, the large, faceless target is big business. Some of this antipathy has its roots in the historic labor union conflicts with big business.

The isolated, dramatic, personal anecdote of some environmental issue is the *smoke* that is fanned into the *flames* of public outcry by media for government to regulate big business, be it industrial manufacturing, mining, oil, home builders or their financiers. With the flames lit, the more public talk there is about the reported environmental issue, the more likely are other accident- monitoring agencies such as police and insurance to collect similar examples of the reported issues that they would have ignored altogether or classified differently prior to the reports. Psychologists call this the "Pygmalion effect." Further, fear mongering relies upon what psychologists refer to as the *availability heuristic*, where people judge the significance of an isolated issue by how readily it comes to mind. When we are presented with a survey that polls the relative significance of an issue, we are likely to give greatest significance to whatever the media emphasizes at the moment because that issue tends to come to mind (57).

Fear mongers make their scares all the more credible by having professional spokespersons or *victim-cum-experts* spread pseudo-scientific or dramatic testimonial information about an environmental issue. Professional narrators also play an important role in transforming implausible environmental threats into the "disaster de jour." The rantings of alarmist newscasters and the glorification of wannabe experts are two tricks that expose the fear mongers manipulative ploy. In addition, the use of tragic anecdotes in place of statistically-tested scientific evidence, and the conflation of isolated incidents into trends or conspiracies are an attempt to exploit deeper cultural anxieties and hatreds, and personal paranoias. The simple answer as to why Americans harbor so many persistent and irrational fears is that immense power and money await those who tap into our moral insecurities and supply us with symbolic substitutes (57).

The US wastes tens of billions of dollars and human resources every year on fear monger-promoted enviro-mythological issues, including research and technology, and on victim compensations for *metaphorical illnesses*. Metaphorical illnesses include Gulf War Syndrome, multiple chemical sensitivity and silicon breast implant disorders where people justify their personal fears, prejudices, hardships and political ideologies in the absence of determinant scientific medical explanations by projecting themselves into a public class of victims (59).

Metaphorical illnesses have become popular in recent personal and class action injury lawsuits. Here, gullible, sympathetic and big business-loathing juries and jurists have evolved a system that perversely exhibits the wealth redistributive characteristics of Marxist socialism. Take for example the hundreds of billions of dollars involved in state health agency *litigation jackpots* won by damage suits against the big tobacco companies. Actual scientific cause and effect of injury for consenting-adult smoker claimant illnesses were never established.

Cause and effect of environmental impacts can be reliably established when the significance of valid scientific data is subjected to statistical analysis—it tells you whether sufficient data are available to determine the cause and effect theory tested. Scientific cause and effect cannot be determined by coincidence, personal anecdotes, opinion polls, circumstantial evidence or preponderance of court evidence which are prevalent legal standards, not scientific conclusions.

A textbook and cinematic example of injury lawsuits foregoing any basic scientific cause and effect findings was portrayed in the 2000 movie "Erin Brockovich" starring actress Julia Roberts. Based on a "true story," the movie advertising line boasts "feisty single mom brings a small town to its feet and brings a big corporation to its knees." This is the oft-heralded environmental *David-vs.-Goliath* theme, where the weak (or defender of the weak) takes on the big bad corporate monster. In this movie, what was billed as victory for innocent rural drinking water consumers over the big greedy corporate giant was in fact an opportunistic trial lawyer's shakedown of a closely-regulated state government utility using evidence that never established any scientific cause and effect between local water consumption and local medical claims. Even if the medical claims had been linked to the public utility as the source of drinking water pollution, it was a failure of the state and local government drinking water regulators not to have identified and abated the reported incidents of water pollution. This was not a case of corporate free enterprise secretly poisoning local yokels for the almighty dollar. The spin of the movie was that big greedy industry should not poison innocent area residents and try to conceal their pollution. Public utilities, such as in the Brockovich movie, are publicly financed and under strict state regulation. They operate as de facto "state industries." Though causation was never traced to the public utility, judges in this class action lawsuit responded both to the fear mongered motivation of moral outrage and the dislike of big business in rendering a multi-million dollar judgement for the plaintiffs and self-righteous trial lawyers. This outcome, as with other recent class action lawsuits such as "big tobacco companies" and "big drug companies," can be referred to as the *Brockovich Syndrome*. Here, corporate America has become a benevolent benefactor for consenting and often unimpaired consumers in the legally-sanctioned jackpot of class action litigation, ensuring literally billions of dollars for future full employment of the predatory plaintiff lawyers. Big business and its consumers are the ultimate losers. Business loses profits and brand reputation for growth, research and development.

Chapter 4

Whose Land

Land is a unique, irreducible, immovable physical asset that confers the most tangible privileges and responsibilities for man's basic claim to the Planet Earth. US land ownership and use is an inalienable right that must be balanced against broader public interests. Democratic free enterprise has private property rights as a founding and sustaining tenet.

Political Lands

Land use, of both public and private property, has become a very controversial environmental issue in the US. Many of the majestic natural symbols of American freedom, such as the scenic mountains, forests, lakes and vast plains of the American West, are public lands owned and managed by the federal government or individual state governments. Half of the land occupied by the 12 westernmost states is actually owned by the Federal Government in distant Washington, D.C. Federal lands comprise 83% of Nevada, 68% of Alaska, 64% of Utah and 44% of California. These lands, totaling approximately one million square miles, are administered by the US Department of Agriculture and US Department of the Interior, or government agencies and bureaus therewithin. These lands cover an area that is approximately 27% of all the US land mass or 9 times the area covered by all US cities, towns, suburbs, factories, offices, roads and highways combined. Much of these lands are national parks, national forests, national game preserves, national sanctuaries, monuments, etc. Most federal lands are managed as "open space" for habitat conservation, wilderness or reclamation with only limited public access for low impact, passive recreational uses. Other federal lands allow private and corporate commercial uses such as grazing, mining and petroleum extractions, logging, lodging, recreations and concessions, with nominal lease, royalty or use fees paid to the federal government. Environmental activists are constantly at odds with these commercial uses on federal lands. Public lands thereby become essentially *political lands*. Unless these public lands can be protected from special interests, both commercial and environmentalist, good management and good scientific findings, regrettably, will play a secondary role in decisions to manage these public natural resources (49).

Private Lands

The concept of private property rights dates back at least 2,000 years to the annals of Roman law. Property rights can be viewed as one of the most fundamental civilizing and motivating features of a free society. Without private property there can be no democracy, no prosperity, no peace and no liberty (20). An 18th Century Colonial American suggested that private property rights are "the guardian of every other right in a free society." Early American property ownership was, along with voting rights, restricted to white men. Over history, with private property rights protection, the exchange of property became increasingly "horizontal" in distribution—from seller to buyer—and decreasingly "vertical"—from father to son through social class or estate lineage. Therefore, property became more and more acquired by the merits of hard work and ingenuity, rather than by inheritance among the privileged aristocracy. Thus, land contracts prevailed over class status in promoting economic opportunity. Wealth was democratized through the commerce of private property rights (13). This commerce of private property rights has evolved to what we now refer to as the real estate business.

Concerning land ownership, the 5th Amendment of the US Constitution states "nor shall private property be taken for public use without just compensation." Today's environmental regulations at local, state and federal government levels flirt with the constitutionally protected property right against "taking" in very oblique, yet insidious ways (25). Environmental regulations that protect wetlands or kangaroo rats or kelp beds or spotted owls can directly take private property rights by government requirements of buffer zones, land set-a-sides or by loss-of-use restrictions.

Environmental regulations can dramatically reduce the value of private property. For example:

- Under the Endangered Species Act (ESA), the federal government can prevent owners from building on their private property, harvesting timber and even walking on land in order to protect endangered animal species. In many cases, the result is not only loss of the use of private property, but extra expenses in delays and construction.

- The current interpretation of provisions in the Clean Water Act allow the EPA and the US Army Corps of Engineers to prevent an owner from building on private property deemed an official "wetland." To obtain permission to build, owners must sometimes pay tens of thousands of dollars in "mitigation" ($50 to $70 thousand per acre of impact), paying for the preservation of another wetland in return for the one they fill in. People have gone to jail for failing to comply with wetland requirements.

- Other laws from protection of beach erosion and requirements for public access have also raised the 5th Amendment takings question.

  Keep in mind that these are **not** situations where a property owner is actually polluting the environment. Rather, these examples illustrate cases where the government requires a private property owner to take on an extra burden that others are not required to shoulder. A property owner is being told to provide an environmental good, often at great personal expense. If he or she were doing something environmentally harmful, that would be different. Environmentalists who oppose government compensation for private property takings sometimes confuse private property uses with pollution, perhaps deliberately (60).

Regarding the taking of private property rights, consider the case of a building permit that conditioned approval of a new home development project on the dedication for, and construction of, a public bike path. The bike path case went to the US Supreme Court in 1994 where the court found that "an unconstitutional taking" of private property rights had occurred. In real estate today, often the property right is compromised for a greater gain in seeking project approvals where environmental impacts must be compensated for, or mitigated. The question of "unconstitutional taking" probably could apply to such mitigation requirements. Court cases have established a useful threshold for "unconstitutional taking" of private property rights to be where 30 to 40 percent of the subject property's value is lost under a government enforcement action (28).

Environmentalists argue that compensating private property owners when government regulations "take" private property by restricting certain uses of the property is tantamount to paying (or bribing) people not to harm the environment. Nothing could be further from the truth. Property rights do not include the right to pollute or harm the environment. Pollution or other environmental impacts upon private property is a *public nuisance* because it infringes upon the property rights of others (52).

Financially strapped municipalities increasingly are funding public improvements through exactions on private development instead of via politically-perilous new taxes. Demands for "on-site" improvements by developers, such as roads within subdivisions, rarely present problems. Exactions for directly related "off-site" improvements, such as the widening of a side road between a highway and a new regional shopping center, often are reasonable. However, cities from Boston to San Francisco are imposing exactions on developers to fund such general public purposes as public housing or job training centers.

Comprehensive data are almost impossible to obtain, since exactions wear many labels, such as "connection," "linkage" and "inspection" fees. Officials often claim not to impose charges at all, and instead hold up projects informally until developers make "voluntary contributions" of funds or services.

The Supreme Court is left to untangle the collision of two basic principles of American law. On one hand, the state has an extensive right, referred to as "police power," to protect the public health, safety and welfare. On the other, the state has a duty under the 5th Amendment Takings Clause not to take private property except upon "just compensation."

The Court noted in 1960 that a primary purpose of the Takings Clause is "to bar Government from forcing some people alone to bear public burdens which, in all fairness and justice, should be borne by the public as a whole." The Court's takings jurisprudence, although unclear, reflects two fundamental concepts. First, that government actions affecting economic rights will be less closely scrutinized by the courts than those affecting fundamental civil rights such as speech, voting, or the rights of minorities. Second is the notion that judicial deference is not unlimited. Some argue that increased protection for private property rights would discourage vital public works. Others say that compensation would force municipalities to take into account the costs of improvements as well as the benefits (67).

Chapter 5

Ecopolitical Abuses

The multiplication of US environmental regulations has become one of redundant and overlapping local, state and federal controls enacted often without the establishment of a scientific cause and effect basis, or economic impact analysis. In ecopolitics, the driving force for new and expanded environmental regulations is political expediency (correctness) rather than good science applied to problem solving. Abuses of the environmental laws have increasingly produced a trial lawyers' "gold mine" in court and private settlements against property owners and business interests. The following are five actual environmental regulatory cases involving ecopolitical abuses by environmentalists and bureaucratic collusion, and trial lawyer exploitations.

Case 1

The US Fish and Wildlife Service forced a new hospital to move its construction site 300 feet in order to avoid and set aside two acres of dune habitat for eight individual flies (flying insects) on the Federal Endangered Species Act list. The protected flies were found near the proposed new hospital construction site. The land set aside for the flies ultimately cost the hospital $3 million and years in construction delays. The same fly species and Federal Regulations delayed a paper-recycling project until the county government could come up with $100 million for a fly habitat land preserve (18).

Case 2

Recently so-called "citizen suits," or threats thereof, brought by tax-exempt environmental groups for alleged violations of the Federal Clean Water Act have become instruments of blackmail or *eco-extortion* against small business operators. In this perverse situation, the tax-payer-subsidized environmental group, ostensibly operating in the public trust and on behalf of public welfare, identifies a business whose public record of compliance with its industrial storm water quality permit is less than perfect. Next, the environmental group notifies the business of its alleged violations and of the group's intent to sue in Federal Court for public damages and legal fees amounting to hundreds of thousands of dollars, with the proverbial offer to settle out of court. The business is left with the burden of proof to defend its compliance record at great time and expense with the uncertain outcome of a Federal Court decision. Often the business will settle out of court with the environmental group by having the US Department of Justice approve the settlement. The settlement money must cover the plaintiff environmental group's legal expenses, and the remainder of the money is distributed at the court's discretion to other non-profit organizations, perhaps to perpetuate and expand this government-sanctioned eco-extortion (15).

Case 3

A private property owner of 300 acres of undeveloped raw land wanted to build 250 new single family homes, including all streets and utilities improvements. The city, in approving the proposed housing development, required the builder to construct roadway improvements on adjacent properties owned and controlled by county, state and federal governments as floodways and public recreation areas. Due to the common, redundant and overlapping environmental regulatory reviews of the city, county, state and federal agencies for the private and public lands involved, the home builder had to gain consensus reviews and approvals from as many as twenty-two regulatory authorities at a cost to the builder of over $4 million dollars and four years of delays to complete the construction of the new homes. These additional environmental reviews and approvals, and public works constructions added significantly to the cost of each new home built, thus adding to the area's home affordability problem. The US Bureau of the Census estimates that every $1,000 increase in home prices excludes 300,000 families from the ability to afford a home (51).

The following is a list of the 22 environmental regulatory agency reviews and approvals needed to build the new homes on private property in this case:

California Environmental Quality Act

California Fish and Game Code

California Endangered Species Act

California Historic Preservation Office

California Regional Water Quality Control Board

California Department of Transportation

County Department of Public Works

County Department of Parks and Recreation

County Department of Regional Planning

County Flood Control District

City Department of Public Works

City District Engineer

Department of Parks and Recreation

US Army Corps of Engineers

US Fish and Wildlife Service

US Environmental Protection Agency

Federal Clean Water Act

National Historic Preservation Act

Federal Clean Air Act

National Environmental Policy Act

Federal Floodplain Management Act

The Vatican

Case 4

Environmentalist litigants are increasingly using the federal and state *False Claims Acts* to impose hefty financial consequences on alleged violators of environmental laws and, in the process reap rich rewards for themselves. The Federal False Claims Act provides that anyone who knowingly submits a "false claim" to the Government is liable for damages and penalties under the laws. Most states also have false claim laws modeled after the Federal Law. Under federal and state laws, private individuals may sue on behalf of the government (when the government has declined to prosecute the action), and share any monetary damage awards with the government. The acts, moreover, can triple any such damage awards.

How can acts designed to deal with "false claims" against the government, like a contractor's submission of a bill to the government for barrels of gunpowder partially filled with sawdust, reach alleged violations of environmental laws? Enter the so-called "reverse false claim," which is the idea that someone who violates an environmental law and fails to report the violation to the government has effectively made a "false claim" to diminish or avoid an obligation to the government. Some recent cases have applied reverse false claim theories against defendants who submit false or incomplete information regarding environmental cleanups, permits and monitoring data.

Yet another court case has concluded that because the purpose of the environmental laws is to compel an environmental cleanup or collect damages related to environmental contamination, the remedies under the environmental laws are different than under the Federal False Claims Act and, thus, they are not mutually exclusive and can be employed simultaneously. Thus, environmental groups can combine false claims litigation with more traditional lawsuits based upon violations of environmental statutes. The use of federal and state false claims actions, coupled with allegations of violations of environmental statues, may subject violators to treble damages in addition to the statutorily prescribed penalties under the environmental statutes (65).

Case 5

Since the 1849 Gold Rush days, the State of California has seen intense competition for land, living space and natural resources. Southern California continues to grow well above the national average in population, with the coincident need for new homes. An area of about 1,400 acres of undeveloped land originally owned by the late industrialist Howard Hughes in coastal Los Angeles has been proposed as a commercial and residential development project, attracting great environmental controversy since the mid-1980s. The single most sensitive environmental feature in the development project has been a wetland area (swamp) on the property known as the Ballona Wetlands, fed by urban runoff drainage waters from the West Los Angeles urban area.

The development project (we will call "Vista") would construct new offices, stores, recreation areas and homes requiring numerous and lengthy, city, county, state and federal government environmental approvals. By law, each of the various environmental approvals require procedural opportunities for "public participation" via open public hearings and comment letters on the environmental merits or problems of the Vista development project. Public participation is built into all US environmental permit and approval decisions at city, county, state and federal levels. Today the requirement of public participation has become a point of access for environmentalist propaganda, political exploitation and exaction of lands and/or money to calm environmentalist objectors. Here is a public forum where literally **any** layman's statements (so-called *interested party* or *stakeholder*) will become part of a project's public record alongside the work of a credentialed scientific expert. Unfortunately, the penalty of perjury does not strictly attach to these public administrative proceedings. This is the perverse process where the theatrics of ecopolitics thrive, often supplanting good scientific evidence in the management of our natural resources, and where erosion of 5th Amendment private property rights is chronic. Here, the results of these administrative proceedings can have the same punitive impact, in costs and delays, of adverse judicial proceedings (lawsuits) upon landowners where ecopolitics is allowed to reign.

By the vehicle of the legal requirement for public participation, environmentalists or other adversarial commercial interests can drive the environmental regulatory process to often stop or reduce the size of legitimate private property developments, and can routinely slow the approvals process and extort lands, costly mitigation measures or cash money by the open and unqualified threat of continued objections and protests. It's political power versus scientific truth at the expense of private property rights and effective management of our nature resources.

In the particular Vista development project case, during over a decade of negotiations with local, state and federal government environmental regulators having jurisdiction over the Ballona Wetlands that are only partially impacted by the proposed project, dozens of established national, international and California environmental groups objected to the project. In addition, a succession of new local environmental groups protested the project for other issues ranging from archeology to zooplankton and everything in between.

Every time one environmental group's issue was settled, another new or offshoot protest group and issue would miraculously appear. Protestors picketed the Vista site and packed public hearings. Numerous lawsuits were threatened and filed. The wetland environmentalists' attacks on the proposed Vista development reached a crescendo after three legendary Hollywood movie and music entertainment moguls planned a new studio within the development. We will refer to the moguls' studio as "DreamTeam." Here is where a little investigation into just who (or what) was behind the insidious objectors revealed not just a competing commercial adversary sponsoring ersatz environmentalist protestors, but something darker in the form of prickly personal revenge inside the entertainment industry.

Interviews with the Vista environmental and public relations attorney yielded some insight into the profiles of the ever-emergent environmentalist protestors. The Vista attorney revealed that certain individuals among the members in the parade of protestor groups were exclusively retained by cash from certain disaffected entertainers to punish DreamTeam by contriving and publicizing environmental issues, intervening to disrupt, confuse and stagnate the approvals process in public hearings, community workshops and lawsuits. Correspondingly, interviews with a Boardmember of the original wetlands environmental group confirmed that some of the protest groups drew contributor money through law firm fronts in order to conceal contributors who were out to get personal and commercial revenge against the DreamTeam moguls.

In addition, the Vista development became such a large target for environmentalist exploitation to attract contributor funds, that the original wetlands groups found themselves competing heatedly with ever-emergent groups for issues, publicity and contributions. It is also instructive to note that in early 1999 the DreamTeam studio pulled out of the Vista development, at which time many of the environmental protest groups faded back into the swamps. The calm following DreamTeam's exit from the Vista project confirms that much of the latent environmentalist attacks were not about saving wetlands, but about punishing DreamTeam. This is but one of many instances where the public environmental approvals process is being used by elitist groups, estranged individuals and commercial competitors for commercial advantage and personal revenge in the guise of championing the public interest and welfare by saving the environment.

Dealing with Environmental Regulators

Should you ever have the daunting task of acquiring environmental approvals for some new venture—commercial, industrial, residential or agricultural—you should be prepared to deal with government agencies that do not share your objectives, your risks or your passions. Given the preceding section's summary examples where ecopolitics played a central role in enabling the redundancy, time and uncertainty of encounters by good faith applicant ventures with local, state and federal government environmental regulators, some general characteristics of environmental regulators may assist in determining the ultimate approvals and success of a proposed venture. Twenty-five years of assisting new ventures in gaining multi-level government environmental regulatory approvals has identified four characteristics of government regulator confidence and competence permutations that can be indicators of the relative success in pursuing their approvals for a proposed venture. As shown in the matrix below, "Regulator Characteristic" 1, 2, 3 and 4 are the four possible combinations of confidence and competence in the environmental regulator, with the corresponding approval "Success Rate" expressed as a percentage.

Regulator Characteristic	Regulator Confident (Yes/No)	Regulator Competent (Yes/No)	Success Rate (%)
1	Yes	Yes	80%
2	No	Yes	80%
3	No	No	50%
4	Yes	No	10%

The theory behind the confidence/competence characteristics is that a "competent regulator" is more useful to a proposed venture than a "confident regulator." The proposition is that you can not make an incompetent regulator competent, but you have an opportunity to instill confidence in the regulator who is not confident. Therefore, Characteristics 1 and 2 carry the same rate of success, 80%. With Characteristic 3, a regulator who is neither confident nor competent, is a toss-up at 50% success rate.

The least success (Characteristic 4 at 10%) involves the confident environmental regulator who is incompetent. Here the environmental regulator may promote a politically-green agenda to stop a proposed venture, or, alternately, under-scrutinize a proposed venture in issuing approvals hastily with a record that will not prevail under administrative and judicial reviews. It is important to understand, in dealing with government regulators, that the regulator has the most secure and well-benefited employment available, without worker competition. The regulator, therefore, has little or no reason to act on any request for approval. You are left with the tenuous task of negotiating approvals with a system that generally has neither the incentives of commercial risk nor professional accountability. This theory holds true as a primary and productive assumption for anyone pursuing environmental approvals with local, state and federal environmental regulators.

Chapter 6

10 Environmental Truths

There are many stunning successes from the environmental controls that were enacted and enforced over the last 30 years in the US. You and I can be quite proud of the successes because we participated in the public, political will that brought environmental regulations. In addition, we have paid its enormous economic costs that are embedded in every product, service and activity we enjoy today. It is estimated that environmental regulations in the US cost consumers between 4% and 5% of our Gross Domestic Product (GDP)—the equivalent of approximately $400 billion on an annual basis (29), or approximately equal to the total annual US Government expenditures on public schools (26). US pollution control costs are roughly apportioned at 3% federal expense, 10% state expense, 13% local expense and 74% private expense (29).

We must attribute a significant part of our present, nearly two decades, of continuous US economic growth and prosperity to the domestic and global assimilation, and market adjustments, to the ubiquitous economic costs of environmental regulations. It has taken a generation of Americans toiling under the economic burden of environmental regulations to reach such prosperity. These successes prove unequivocally that environmental protection can only be supported in democratic and prosperous economies such as the US. It all started in America. Environmental protection can not be a priority in third world or developing countries. They simply can't afford it as a priority.

The following 10 referenced truths define the **true** state of today's environment:

Truth 1

World population has more than doubled since 1950, while food supplies have more than tripled. We are feeding today's world population on essentially the same 37% of the planet's land area as we did in 1950. Higher crop yields per acre due to the use of herbicides, pesticides and genetic engineering have saved more than 10 billion acres of wildlife habitat (10).

Truth 2

Since 1970, US population has grown by 30%, GDP by 110% and motor vehicle miles driven by 130%. Since 1970, US aggregate air pollution emissions have decreased by more than 30% (61). Between 1990 and 1999, common air pollutants in the US such as sulfur dioxide (SO2) levels are down 36%, carbon monoxide (CO) down 36%, smoke, soot and particulate matter have been reduced 18%, lead down 60%, nitrogen dioxide down 10% and smog down 4% (70). Once the nation's worst, Los Angeles smog has been cut in half while the city's motor vehicle numbers increased by 65% since 1975 (31). Since the mid-20th Century, Southern California's population has grown from 4.8 to 15 million people (more than tripled), while smog has been reduced by 75% (5). Since 1970, levels of air borne lead dropped by 98%, CO by 24%, SO2 by 30%, soot and particulate pollutants by 78% (29). In 1999, the Los Angeles area had a rate of lung cancer that is significantly lower than both the California State and US national lung cancer rates (71).

Truth 3

Only 7 of the approximately 1,500 species listed for the protection under the Endangered Species Act since 1973 have become extinct, and about 25 have recovered to sustainable populations (11). Some of these include the Bald eagle, Peregrine falcon, American alligator, Timber wolf and others. No species of primate (apes and monkeys) has gone extinct in the 20th Century (30). The US Endangered Species Act costs $227 million per year to operate (16).

Truth 4

The total amount of large-tree standing timber in the US has increased by 30% since 1950 (56). US forestlands covered 732 million acres in 1920; today US forestlands cover 737 million acres—a 5 million-acre increase today (14). By comparison, all US cities, towns, suburbs, factories, offices, roads and highways combined cover less than 60 million acres in the US (13)—an area about the size of the state of Oregon or 3% of the US land area (26).

Truth 5

Forest lands in Europe have expanded from 361 million acres to 482 million acres, increasing 34% over the last 40 years (14). The South American Amazon rainforest is 90% intact as compared to 30 years ago when measurements began. Worldwide, rainforests today cover more of the earth than they have for 10 thousand years (35).

Truth 6

The US banned ozone-depleting chemicals, chlorofluoro-carbons (CFC's), in 1978, and the ultraviolet protection of the stratospheric ozone layer is now recovering (9).

Truth 7

Of the nearly 1,400 US hazardous waste Superfund sites being cleaned up, ½ have been completed, and all will be cleaned up within 5 to 7 years (9). The amount of widely used toxic chemicals released in the US has been cut in half since 1988 (6). The EPA reports that toxic chemical releases from US manufacturing declined by 90 million pounds between 1997 and 1998 (45). The amount of urban garbage being disposed of in sanitary landfills has been reduced by 20-30% since 1980 due to recycling and waste diversions (7).

Truth 8

Exxon has spent $2 billion dollars to clean up Alaska's Prince William Sound since the infamous quarter-of-a-million-barrel Exxon Valdez oil spill ten years ago. The wildlife and habitat in the Sound have largely recovered (3). The prestigious journal, "Scientific American" has concluded that in areas where the oil was left to the natural cleansing processes of wind, weather, waves, biodegradation and sunlight, recovery of wildlife habitat is better than those area shorelines that were effectively sterilized by high-pressure steam cleanings advocated by over-zealous environmentalists (13).

The Trans-Alaska Pipeline was proposed in 1968 to convey oil from reserves in Prudhoe Bay on the northern coast of Alaska 800 miles to the port of Valdez on Alaska's southern coast. Construction of the pipeline was claimed by environmentalists to be destructive to wildlife and their habitats, including herds of elk, caribou, bald eagles and arctic foxes among others. Environmentalists delayed the pipeline for several years, adding tens of millions of dollars to the cost of the pipeline (38). Environmental studies during the planning of the oil pipeline determined that 3,000 caribou lived within, and may be impacted by, construction and operation of the pipeline. Today, 30 years after the pipeline was built and placed in service, studies indicate that more than 20,000 caribou live in the area of the pipeline (39). Since 1977, these

Alaskan oil fields have provided more than 12.5 billion barrels of oil, or 20% of US production (40).

Truth 9

The numbers of US lakes, rivers and streams that are deemed to have water quality that is safe for fishing and swimming have doubled since 1972 (29). There has been no net loss of US wetlands since 1980 (32).

Truth 10

The EPA's own records indicate that all forms of environmental pollution taken together cause just 1-3% of all cancers, while 30-35 % of cancers are attributable to heredity and personal indulgences of diet and substance abuse. Recently, the Harvard Center for Risk Analysis showed that federal government pollution controls cost $7.6 million to save a single year of one's life, while medical care to save a single year of one's life costs only $19,000 (1). Life expectancy worldwide has risen from 47 years to more than 64 years during the 20th Century—this is the greatest advance in overall human welfare in history (1).

Chapter 7

Future Environmental Concerns

The US environmental movement and regulatory system reached its full, practical expression in the 20th Century. In the 21st Century the US environmental movement and regulatory system should fully evolve into "state primacy" (i.e., state and local level regulatory responsibility) in enforcement of environmental regulations. The movement and its environmentalists must concede that virtually every thing "regulatable" is regulated, and that the energy of environmental activism should 1) be directed globally to underdeveloped and developing countries or 2) be directed to improvements in US education and welfare issues with moderate ideological foundations and proven successes.

The most important future environmental concerns should be the export and application of US pollution control technologies and 21st Century electric power generation innovations to underdeveloped and developing countries. The US Government, in partnership with commercial technology companies, should adopt and fully fund research to invent the next electric power generation systems with the resolve, vigor and national pride of the lunar landing initiative of the 1960s. The US and other advanced countries are still using 19th Century electric power generation technologies in the 21st Century. In the US, only 4% of the nation's electricity is generated by oil, compared with 52% by coal, 15% by natural gas, 19% by nuclear reactors and 10% by hydroelectric plants (62). The next electric power generation technologies should be open to the consideration of new nuclear power technologies. Underdeveloped and developing nations, with explosive population growth, desperately need new, cleaner, more efficient electric power generation systems.

Other 21st Century environmental technology concerns and innovations may include (60):

Agrogenetics

Genetic engineering and plant crop manipulation will reduce agricultural land use areas and impacts upon sensitive environments. Crops should be developed to require fewer amounts of pesticides with greater pest resistance. Crops should be engineered to require less fertilizer and water while providing higher yields.

Smart water treatment

Smart membranes and filters should be developed to improve water treatment quantity and quality at sewage plants and municipal drinking water supplies.

Renewable energy storage

Improved electric power storage must be developed to increase the viability of electricity generated from feasible solar and wind power. Renewable energy sources can reduce reliance upon imported oil.

Microtechnology

Residential and commercial interiors should be heated and cooled more efficiently by micro heat pumps and air conditioners. Micro-scale chemical plants development would produce industrial chemicals on demand, reducing storage requirements, wastes, transportation hazards, and improving economy of scale.

Paperless society

Innovative displays, wireless communications and customized web publications will continue to reduce the use of paper. Advanced display systems may imitate paper in their flexibility and portability.

Molecular design

Molecular design of catalysts could make chemical reactions and industrial processing so precise that little or no wastes are produced. Sensors designed at the molecular level could monitor material and chemical manufacturing.

Bioprocessing

Microorganisms and plant species should be developed to grow biorational chemical and biological products such as drugs, proteins and enzymes. Producing chemical feedstocks, fuels and pharmaceuticals in this manner could be cost-effective with minimal waste generation.

Real-time environmental sensors

Supermarkets should develop sensors to detect *E. coli* and other dangerous pathogens in food supplies. Workplace air quality could be monitored to prevent "sick building syndrome." Other preventative health benefits could include monitoring the environment on airplanes, preventing infections at hospitals and in drinking water supplies.

Enviromanufacturing and recycling

Plastics, paper, cars and computers will be more recyclable or biodegradable. Dry cleaning with liquid carbon dioxide would minimize or eliminate hazardous waste.

Lightweight cars

Squeezing every ounce possible out of cars will mean a family sedan that gets at least 80 miles per gallon of gas and generates less pollution. Lighter cars will be built with less steel and more lightweight aluminum, magnesium, titanium and composites. Advanced metal-forming techniques will provide precisely the strength needed at every point of construction.

Environmental education

Understanding complex living systems adequately to effectively manage our activities and engineer appropriate solutions are on-going challenges. Complex living (ecological) systems include: the urban atmosphere; lakes, oceans and coastal waters; and underground (subsurface) environments. Although these systems vary in size and character, each system is accountable to biological, physical and chemical interactions that govern reactions and transport processes within the system. The engineers and scientists of the future must be able to apply analytical and problem-solving skills to understand and predict these complex environmental systems, and be able to implement strategies for resource management and repair. The need for multi-disciplinary scientists is an absolute requirement in understanding these complex systems.

The American public should be proud of the vast improvements made to the health of their environment in the last 30 years. The environmental fad of 1960s and 70s appears to have lasted longer than many would have predicted. Indeed, the US environmental consciousness continues to be raised and an improvement in the overall quality of life for the masses is a goal for all environmental engineers and scientists. Scientific education is, and shall always be, the cornerstone for prosperity, cultural advancement and prudent natural resource management in free democratic societies.

So there you have it, environmental issues presciently recognized by President Teddy Roosevelt in the early 20th Century led to land conservation measures to control the irresponsible use of our natural resources (environment). In the last half of the 20th Century, US governmental agencies at all levels imposed vast regulatory systems to control virtually every human, commercial and industrial activity in the US—often without conclusive scientific evidence of any significant impact of the activities upon ecology or human health. Environmental issues where elevated from local to national significance, leading to national environmental regulations. In the last decade of the 20th Century, environmentalists attempted to globalize environmental regulations via the vehicle of the global warming issue. The globalization did not work, nor did a critical mass of worldwide political power materialize. In the US, with the completion of the national environmental regulatory system, the focus of environmental issues will shift to state and local level solutions in the 21st Century.

Environmental regulations have spawned not only the most comprehensive (and some would argue the most efficacious) governmental involvement in every endeavor of American Life and commerce, but also have propagated a succession of environmentally co-dependent enterprises; from the growth of multi-level government regulators and their agencies, to academic curricula in environmental studies, to environmentalist activism, to the profusion of technical and legal consultants, to the *green market place*, and ecotourism.

You cannot name a single product, service or activity of your daily American Life that is not subject to some government environmental regulation. O.K., maybe the internet, *web* and *.com* services are largely excluded from environmental regulations. Perhaps the lack of direct environmental regulation is a factor in the astonishing growth of internet businesses.

So, maybe now as we begin the 21st Century, we can concede that the tenuous and invasive network of US environmental regulations imposed to protect us and our natural resources is comprehensive, complete and successful if only by the obvious measure of its redundant and excessive scope and costs, and itself, ironically, now producing a negative impact in the form of uncontrolled government. Sound environmental policy must rest on the firm foundation of sound current science. We cannot solve problems of basic human needs without stable, growing economies. Democratic free enterprise and increased global economic integration provide the only proven opportunity to improve the condition of all laborers, make economies more efficient and protect the environment.

Environmental issues and solutions will remain forever a uniquely 20th Century American phenomenon with profound cultural and economic impacts for as long as democratic societies are free to prosper.

Bibliography

1. *Los Angeles Times*, March 28, 1999.

2. *The Wall Street Journal*, January 3, 2000.

3. *Exxon Lamp*, Winter 2000.

4. US EPA History Office, Budget Division [epa.gov.com.], February 25, 2000.

5. *AQMD Advisor*, South Coast Air Quality Management District, Diamond Bar, CA, November 1999.

6. *The Wall Street Journal*, May 21, 1997.

7. *The New Yorker*, April 10, 1995.

8. *Pollution Engineering*, January 2000.

9. *Pollution Engineering*, December 1999.

10. *Saving the Planet with Pesticides and Plastic: The Environmental Triumph of High-Yield Farming*, by Dennis Avery, Hudson Institute Press, 1995.

11. *The Wall Street Journal*, June 30, 1995.

12. *The Wall Street Journal*, May 18, 1999.

13. *Hard Green*, by Peter Huber, Basic Books, December 1999.

14. *Readers Digest*, December 29, 1997.

15. *Los Angeles Business Journal*, January 31, 2000.

16. *The Wall Street Journal*, March 2, 1998.

17. *The Wall Street Journal*, August 31, 1999.

18. *Readers Digest*, January 28, 2000.

19. *The Wall Street Journal*, August 22, 1999.

20. *The Noblest Triumph: Property and Prosperity Through the Ages*, by Tom Bethell, St. Martin's Press, December 1999.

21. *Outside Magazine*, March 1994.

22. *ECON*, February 1995.

23. *Gilder Technology Report*, George Gilder, May 6, 1999.

24. *Atlas Economic Research Foundation*, Alejandro Chatuen, December 6, 1996.

25. *The Economist*, April 6, 1996.

26. *The World Almanac 2000*, World Almanac Books, 1999.

27. *Environmental Protection*, December 1999.

28. *Environmental Protection*, June 1999.

29. *Investor's Business Daily*, April 22, 1994.

30. *The Wall Street Journal*, May 9, 2000.

31. Pacific Research Institute, Greg Easterbrook, April 20, 2000.

32. *Industrial Wastewater*, July/August 1997.

33. Hudson Institute, Michael Fumento, April 2, 1999.

34. American Council on Science and Health, Elizabeth M. Whelan, July 27, 1999.

35. *Tropical Rain Forest: A Political Ecology of Hegemonic Mythmaking*, Philip Stott, Coronet Books, 1999.

36. *The Dictionary of Ecology and Environmental Science*, by Henry W. Art, Henry Holt and Co., 1993.

37. *Websters New Collegiate Dictionary*, Merriam-Webster, Inc. Publishers, 1991.

38. *The Dictionary of Cultural History*, Houghton Mifflin Co., 1988.

39. Former Alaska Governor Walter Hickel, June 21, 2000.

40. *The Economist*, July 8, 2000.

41. *The Economist*, July 15, 2000.

42. *The Wall Street Journal*, July 25, 2000.

43. *California Environmental Law Reporter*, July 2000.

44. *Water Engineering & Management*, June 2000.

45. *Pollution Engineering*, July 2000.

46. *Ecopsychology: Restoring the Earth, Healing the Mind*, Sierra Club Books, May 1995.

47. *Black's Law Dictionary*, Sixth Edition, West Publishing Co., 1990.

48. *Ecology Pollution Environment*, W.B. Saunders Co., 1972.

49. *The Economist*, November 4, 1995.

50. *The Wall Street Journal*, August 16, 2000.

51. *The Wall Street Journal*, August 11, 2000.

52. Competitive Enterprise Institute, Jonathan Tolman, January 16, 1995.

53. *Water Engineering & Management*, August 2000.

54. *Environmental Protection*, May 2000.

55. *Environmental Protection*, July 2000.

56. *The American Spectator*, April 2000.

57. *The Culture of Fear*, Barry Glassner, Basic Books, 1999.

58. *Risk and Culture*, University of California Press, 1982.

59. *Illness as Metaphor*, Farrar, Straus & Giroux, 1989.

60. *Environmental Protection*, June 1999.

61. *Pollution Engineering*, September 2000.

62. *The Power Report*, Gider Technology, February 2000.

63. *The Wall Street Journal*, March 15, 2000.

64. The Long March: How the Cultural Revolution of the 1960s Changed America, Roger Kimball, Encounter Books, 2000.

65. *Washburn Briscoe McCarthy Newsletter*, June 1, 2000.

66. *The Wall Street Journal*, April 17, 2000.

67. *The Wall Street Journal*, March 2, 1994.

68. *Earth in the Balance*, Al Gore, Plume Books, 1992.

69. *The Prince of Tennessee*, Maraniss and Nakashima, Simon & Schuster, 2000.

70. *Environmental Protection*, October 2000.

71. HCIA—Sachs, Dennis Dunn, October 18, 2000.

72. *Environmental Law*, North American International, 1972.

Index

K

L

M

O

oil, 32, 37, 43, 72-73, 77-78, 90

Our Earth, Ourselves, 39

ozone, 33-34, 37, 39, 41, 71

P

PAC, xvi

Paperless society, 79

passivity, 15

perceptions, 9

pesticides, 19, 33, 70, 78, 83

Pinchott, Gifford, xx

Planet Earth, 31-32, 49

Political Lands, 49-50

political left, x, 33, 43

political right, x

politics, 10, 23, 31

policy, 3-6, 8, 21, 23, 31, 60, 82

pollutant, 7, 37

pollution, 6, 10-11, 27, 34, 36-39, 41, 45, 52, 69-70, 73, 77, 80, 83, 85-86, 89

population, 16, 18-20, 32, 42, 61, 70, 77

V

Vietnam War, 8

virtue, 9, 16

W

water, 4, 9, 17, 19, 34-39, 45, 51, 58-60, 73, 78-79, 85-86, 89

water pollution, 36, 39, 45

water quality, 35, 58-59, 73

water supply, 37-39, 89

welfare, 25, 53, 58, 64, 73, 77

wetland, 51, 62-63

WHOSE LAND, 49, 51, 53

wildlife, 6-7, 22-25, 57, 60, 70, 72

Wilderness Society, xviii, 22

world war, 20

World Wildlife Fund, 22, 24

X

Y

Z

zero sum, 31

zooplankton, 63

About the Author

The Author, Paul Taylor was born and raised in a rural community nestled in the foothills of the Blue Ridge Mountains in northeast Alabama. His childhood was a dawn-to-dusk exploration of the outside physical world—the woods, lakes and animals of the temperate pine forests.

In college, he studied biology, chemistry and the aquatic environments on the coast of the Gulf of Mexico. As a college senior, Mr. Taylor spent a summer at a marine sciences field research institute located on the Gulf Coast. Here, coastal ecosystems were investigated, including marine fisheries, and river and salt marsh environments. Somehow through this summer of constant mosquito swarms, endless days knee-deep surveying in smelly salt marshes or parched on some outer island, the Author gained an insatiable curiosity for ecological systems. Later at graduate school in New Orleans, Mr. Taylor studied environmental sciences, focusing on water resources and pollution control; and was the first student to cross-register into the university law school to take environmental law courses. The Author worked his way through graduate school in a medical research laboratory that was awarded the *Nobel Prize in Medicine* in 1977. He also supplemented his income as a model and actor, joining *The Screen Actors Guild* in 1978.

Hired into environmental consulting upon graduation, Mr. Taylor was involved with public works projects of highway, water supply, flood control and waste management impact studies. At another consulting firm, he advised on large river and coastal industrial manufacturing, mining and oil projects in the US, South and Central America, Africa and the Pacific Rim.

Later moving to New York City and then Los Angeles, Mr. Taylor established, and continues, a private environmental science and regulatory consulting practice serving land development, waste management, industrial and attorney clients. Mr. Taylor is a court-qualified expert witness and Registered Environmental Assessor, and has a speaker series based on this book.

www.ingramcontent.com/pod-product-compliance
Lightning Source LLC
Chambersburg PA
CBHW020256290526
45784CB00003B/1271